THE STICKER TREASURY OF

WOODLAND ADVENTURES

Smith Street Books

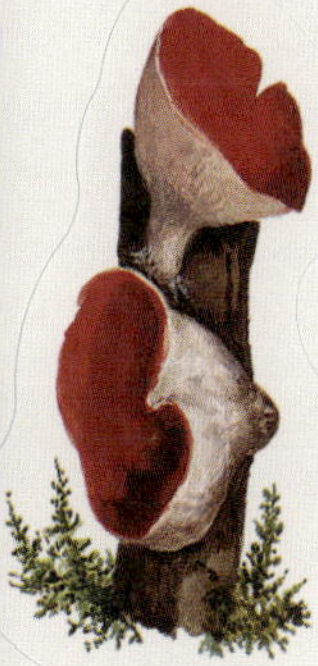

OEWhitney
L. Prang & Co. Boston

RACCOON.

RACCOON.

BADGER

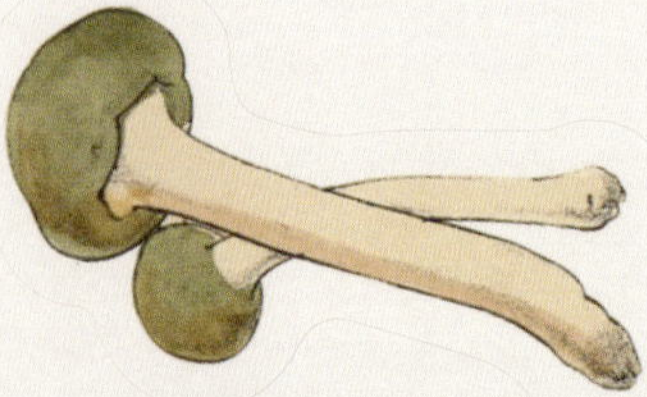

SKUNK.

TEETH OF WEASEL
FOOT OF WEASEL

BEAVER

BEAVER.

C.F.NEWALL

PORCUPINE.

PORCUPINE, CRESTED.

Opossum

OPOSSUM.

A.J.
D.G.

RED FOX.

Howitt

WOLF

月百姿
むさしの月

THE END

GRIZZLY BEAR

G.Berger

SI DIEV LE VEVLT
SANS DIEV NE PEVX

WARWICK GOBLE

VIRGINIAN DEER

常信筆

Mr. Yam

Published in 2025 by Smith Street Books
Naarm (Melbourne) | Australia
smithstreetbooks.com

Distributed outside of ANZ, North & Latin America by
Thames & Hudson Ltd., 6–24 Britannia Street, London, WC1X 9JD
thamesandhudson.com

EU Authorised Representative: Interart S.A.R.L.
19 rue Charles Auray, 93500 Pantin, Paris, France
productsafety@thameshudson.co.uk; www.interart.fr

ISBN: 978-1-9232-3938-8

Smith Street Books respectfully acknowledges the Wurundjeri People of the Kulin Nation, who are the Traditional Owners of the land on which we work, and we pay our respects to their Elders past and present.

Publisher: Hannah Koelmeyer
Editor: Avery Hayes
Design: Holly Dunn
Image research and layout: Heather Menzies, Studio31 Graphics
Production manager: Aisling Coughlan

Printed & bound in China by Asia Pacific Offset Ltd.

Book 406

10 9 8 7 6 5 4 3 2

MIX
Paper | Supporting responsible forestry
FSC® C012521
FSC www.fsc.org